**COUNTDOWN TO SPACE**

# VENUS—
# THE SECOND PLANET

Michael D. Cole

**Series Advisors:**
Marianne J. Dyson
Former NASA Flight Controller
and
Gregory L. Vogt, Ed. D.
NASA Aerospace Educational Specialist

**Enslow Publishers, Inc.**

40 Industrial Road               PO Box 38
Box 398                          Aldershot
Berkeley Heights, NJ  07922   Hants GU12 6BP
USA                                      UK

http://www.enslow.com

**Library of Congress Cataloging-in-Publication Data**

Cole, Michael D.
    Venus : the second planet / Michael D. Cole.
        p. cm. — (Countdown to space)
    Includes bibliographical references (p.   ) and index.
    Summary: Describes the composition of Venus from its core to its
atmosphere and explains how Venus is incapable of supporting life even
though it appears to be similar to Earth.
   ISBN 0-7660-1509-2
   1. Venus (Planet)—Juvenile literature. [1. Venus (Planet)]  I. Title.
II. Series.
QB621 .C67 2001
523.42—dc21
                                               00-008730

Printed in the United States of America

10 9 8 7 6 5 4 3 2 1

**To Our Readers:**
All Internet addresses in this book were active and appropriate when we
went to press. Any comments or suggestions can be sent by e-mail to
Comments@enslow.com or to the address on the back cover.

**Illustration Credits:** © Corel Corporation, pp. 12, 39; Enslow Publishers,
Inc., p. 13; National Aeronautics and Space Administration (NASA), pp. 4,
6, 7, 8, 14, 16, 18, 21, 22–23, 25, 28, 30, 32, 33, 35, 41.

**Cover Illustration:** NASA (foreground); Raghvendra Sahai and John
Trauger (JPL), the WFPC2 science team, NASA, and AURA/STScI
(background).

# CONTENTS

*The Hubble Space Telescope took this ultraviolet-light image of Venus. Venus is covered with clouds of sulfuric acid, rather than the water vapor clouds found on Earth.*

# 1

# Peeking Through the Clouds

About 30 million miles from Earth, the robotic spacecraft *Magellan* was in orbit around the planet Venus. Since the 1960s, scientists had sent spacecraft to the planet to study its atmosphere and surface. *Magellan* arrived at the planet in 1990. The spacecraft's mission was to make a detailed map of the planet. But *Magellan* could not make a map of Venus simply by taking pictures of its surface.

Below the spacecraft, the planet was covered by a thick blanket of clouds that completely hid Venus's surface. Neither astronomers nor spacecraft could see the planet's surface from space. Instead, *Magellan* beamed radar, a special form of radio waves, down through the clouds to Venus's surface. The spacecraft then measured the waves as they returned through the planet's atmosphere. Measuring how the waves bounced back up to *Magellan* created a computerized picture of what the surface looked like.

The spacecraft had been beaming and measuring the waves during hundreds of orbits around the planet. After each mapping pass over Venus, the spacecraft radioed the electronic information it had collected back to Earth. At the Jet Propulsion Laboratory (JPL) in Pasadena, California, scientists who were studying the images found volcanoes all over the planet. Volcanoes had been detected before by other spacecraft, but until the detailed

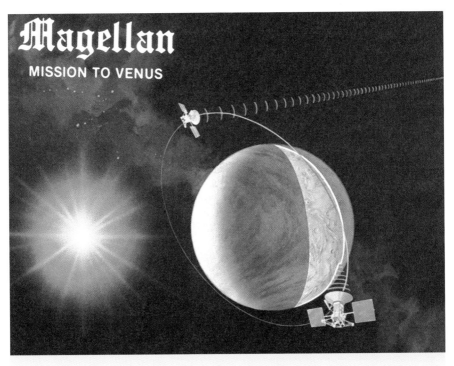

*The* Magellan *spacecraft used radar to map the surface of Venus. Then it sent the information to Earth.*

*This is a computer-simulated view of the surface of Venus, based on images recorded by spacecraft. Gula Mons is a 2-mile- (3-kilometer-) high volcano on Venus.*

mapping by *Magellan*, scientists had not known there were so many. They hoped to find evidence of an actively erupting volcano.

"Just as on Earth, it is very likely that somewhere on the planet a volcano is erupting," said *Magellan* project scientist Dr. Steve Saunders. "The problem is to find it."[1]

The images from *Magellan* showed scientists evidence of a great deal of volcanic activity in Venus's past. Lava flows and wind erosion had shaped the planet's surface, just as lava flows and wind and rain erosion have shaped the surface of Earth. It appeared to

Saunders and other scientists that Venus's recent past had been very violent.

"These images will form the basis for all future scientific studies of Earth's sister planet," Saunders said, "and will provide the necessary maps for all future Venus missions."[2]

In the first forty years of spaceflight, Venus was visited by spacecraft more often than any other planet.

*Maat Mons is a six-mile-high volcano on Venus. This image was made using data from* Magellan *and color images from Venera spacecraft.*

Until the use of radar, its thick atmosphere had kept scientists from answering many important questions about the planet. But modern scientists were only the latest people to wonder about the mysteries of Venus. Humans had been looking with curiosity to our neighbor in space since the dawn of civilization.

# 2

# Goddess of Love, Planet of Heat

Easily seen in dark skies with the naked eye, Venus is an unmistakable planet. When visible, it is the brightest object in the sky after the Sun and the Moon. It is not surprising that it caught the attention of ancient sky watchers.

## Ancient Myths

As far back as 1800 B.C., the Babylonians associated the planet with Ishtar, their goddess of love and war. Almost certainly the beauty of the bright planet caused it to be regarded as a goddess of love by later cultures. The ancient Chinese called it Tai-pe, or the "Beautiful White One."

The Mayan civilization of Mexico based their calendar around the movement of Venus. The 584-day calendar followed the period of time that it took Venus to return to the same position in the sky. The Maya called Venus the Ancient Star and sacrificed their enemies to it when it reappeared after weeks of being lost in the Sun's glare.

"When Venus made its reappearance in the east," one historian wrote, "the Maya sacrificed captives in its honor, offering blood, flipping it with their fingers toward the planet."[1]

Other ancient peoples of Mexico feared the appearance of Venus. When it became visible in the morning or evening sky, they would go inside and close their doors and windows. They thought this would protect them from its supposedly harmful rays.

Like the Babylonians and Chinese, the ancient Greeks associated Venus with beauty. They named it after their goddess of love, Aphrodite.[2] Finally the Romans named the planet Venus after their own mythical love goddess. Since then the planet has been called Venus.

Planets were not merely objects of myth in ancient times. Early astronomers studied the positions of the planets and measured their movements through the sky. Telescopes were not yet invented, so the astronomers used other tools and devices to measure and record the movements of the planets.

## The Planet's Orbit

Over many centuries, astronomers learned that Earth and the other planets orbited the Sun. By using early

*The Romans named the planet Venus after their goddess of love.*

telescopes, scientists such as Galileo Galilei saw the shape of Venus slowly going through phases like the Moon. Over several months, the planet would change from the shape of a crescent, to a half circle, and later to a complete disk.

The reason for this, they discovered, is that Venus's orbit is closer to the Sun than Earth's orbit. Because Earth orbits at a distance outside the orbit of Venus, there are times when Venus passes between Earth and the Sun. Only the part of Venus facing the Sun is lit. Just like the Moon is dark when it is between Earth and the Sun (new Moon), Venus appears dark to us when it is between Earth and the Sun.

## Earth's Sister Planet

As the techniques for measuring the movements and distances of the planets improved, astronomers learned more about the planet Venus. They learned that Venus is the nearest planet to Earth. It is also about the same size as Earth. Earth is about 8,000 miles (13,000 kilometers) wide, while Venus is about 7,500 miles (12,000 kilometers) wide. Venus is often called Earth's "sister planet" because of its closeness and similar size.

But the similarities end there. Its ancient associations with goddesses and beauty led people to imagine that Venus might be a place of waterfalls and gardens—a paradise in the heavens. But Venus is definitely no paradise. In fact, it is much more like an inferno.

*When Venus passes between Earth and the Sun, the bright side of Venus cannot be seen from Earth.*

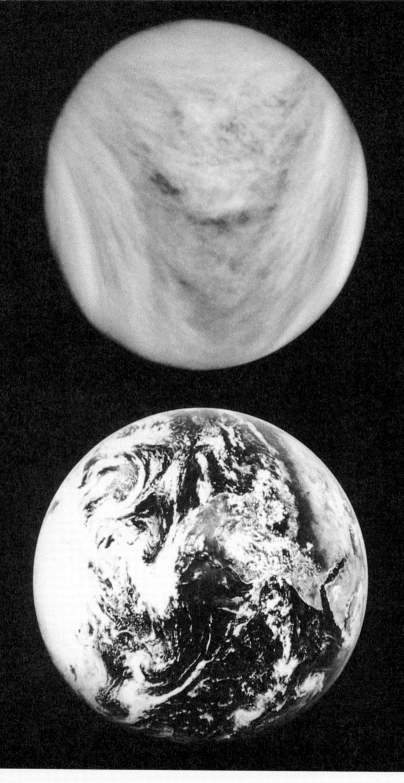

*Venus (top) and Earth are almost the same size. They are commonly called "sister planets."*

### Atmosphere of Heat and Pressure

Although Mercury is the closest planet to the Sun, Venus is the hottest world in the solar system. Mercury orbits at about 36 million miles (58 million kilometers) from the Sun, while Venus orbits the Sun at a distance of about 67 million miles (108 million kilometers). But Venus is completely covered with a thick atmosphere of clouds, made up almost completely of carbon dioxide. Like the glass of a greenhouse, the carbon dioxide allows the Sun's heat to penetrate to the planet's surface but it does not allow all of the heat to escape. Because of this greenhouse effect, the heat builds up. The temperature on the surface of Venus is as high as 900° F (482° C).[3] Such a temperature is hot enough to melt your school desk and reduce your family's car to a puddle of molten metal.

How were such hot conditions created on a planet nearly the size of Earth, at nearly the same distance from the Sun? Some scientists believe that Venus was once very much like Earth, with vast water oceans like the ones on our own planet. The oceans were formed on Venus just as they were on Earth. Gases were spewed out through the surface from the planet's core of molten rock over millions of years.

At this time in the early solar system, the Sun was cooler and dimmer than it is today. As the Sun grew hotter, scientists believe that Venus's oceans began to evaporate into the atmosphere. The evaporation took

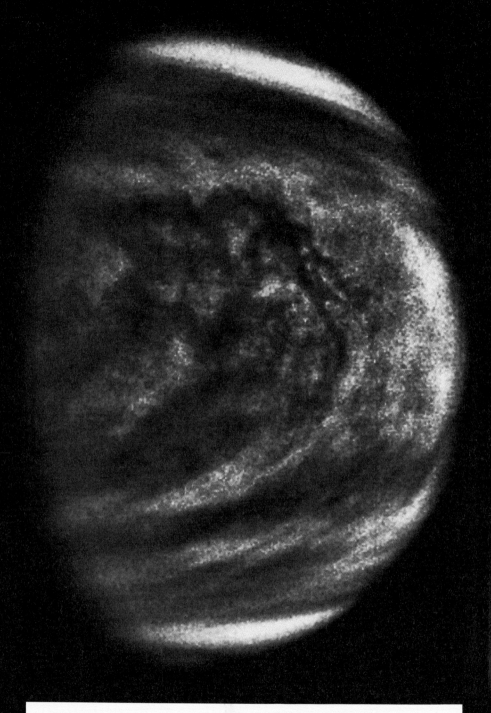

Scientists believe that the Sun's heat caused the oceans on Venus to evaporate. This made the atmosphere very dense. This color-enhanced image of Venus shows the sulfuric acid clouds about thirty to forty miles above Venus.

millions of years and made the planet's atmosphere very dense. The temperature slowly climbed to its current intensity.[4]

Not only is Venus too hot for humans to survive in, its atmosphere is deadly to human life. The atmospheric pressure on Venus is ninety-five times greater than the pressure on Earth. Atmospheric pressure is the weight of the atmosphere pressing down on our bodies. When we are at sea level, every square inch of our bodies is experiencing about fifteen pounds of atmospheric pressure. On Venus, every square inch of our bodies would experience more than thirteen hundred pounds of atmospheric pressure. The pressure on the surface of Venus is equal to the pressure you would feel at almost three thousand feet below one of Earth's oceans. Such a depth is twenty-eight times deeper than any scuba diver could ever go.

Venus's atmosphere is made of 98 percent carbon dioxide and 2 percent other gases. Carbon dioxide is poisonous to humans—we exhale it and depend on plants to soak it up. With so much carbon dioxide in Venus's atmosphere, humans could not survive there.

In addition to the intense pressure and lack of oxygen, Venus has other atmospheric dangers. There are cloud layers of pure sulfuric acid near the surface of the planet. Sulfuric acid quickly corrodes and burns through many substances on Earth, including human skin. Small amounts of sulfuric acid are released into the air when we cut an onion, causing our eyes to produce tears. There are whole clouds of this nasty stuff on Venus. Humans who breathed such a poisonous atmosphere would be killed.

### North and South on Venus

Venus's heat and poisonous atmosphere are not the only strange qualities it has. The planet is also upside down!

Most planets in our solar system orbit the Sun with

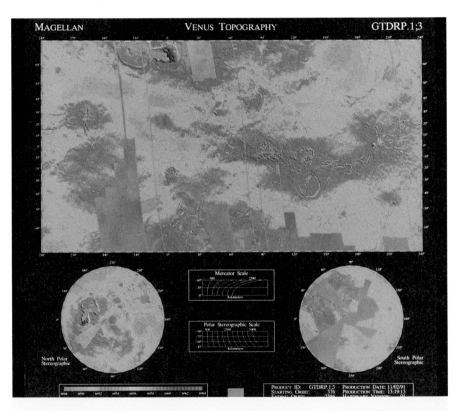

*The data from Magellan were used to make this map of Venus. The north and south poles of Venus are included in the map. Venus's north points in the opposite direction from that of the other planets.*

their magnetic north (the direction pointed to by a magnetic compass) pointed in the same general direction. This northern direction is considered to be "above" the level plane of space in which all the planets orbit the Sun. In other words, north is generally pointing up and south is generally pointing down on most planets in our solar system. But on Venus, north is pointed almost straight down. However, no planet can really be considered "upside down" in space. Venus's north is simply pointing in the opposite direction from that of the other planets.

All planets rotate on their axis, like a basketball spinning on your fingertip. Earth and most other planets rotate from west to east. It appears from our perspective on Earth that the Sun is rising in the east and later setting in the west. Venus's so-called upside-down orientation means that Venus, unlike any other planet in the solar system, rotates on its axis in the opposite direction from Earth. Instead of the Sun appearing to rise in the east and set in the west, the Sun on Venus rises in the west and sets in the east.[5]

In the 1950s, astronomers began taking special photographs of Venus using ultraviolet light, which revealed cloud patterns in the planet's atmosphere. The first ultraviolet images of Venus, taken in 1957, showed astronomers that the clouds were rotating around the planet in the opposite direction that other planets rotate. The images were the first strong evidence that the planet

itself rotated in the opposite direction of other planets in the solar system.[6]

## A Year on Venus

Venus also has a shorter year than Earth. While Earth takes 365 days to orbit the Sun, Venus completes one orbit around the Sun in 225 Earth days. But it takes Venus 243 days to complete one rotation on its axis. One rotation on its axis takes eighteen days longer than its orbit around the Sun. That means Venus's day is longer than its year!

## Studying Venus

While astronomers and scientists continued their Earthbound study of Venus, the space age had begun.

*Scientists were busy in 1969 with the launch of* Apollo 11 *that landed the first humans on the Moon. They were also continuing to study Venus.*

Russia had launched *Sputnik,* the world's first artificial satellite, into orbit in 1957. By 1961, humans had gone into space for the first time, and other robotic spacecraft had traveled to the Moon.

In the 1960s, astronomers studied Venus using radar. By aiming radio waves at Venus and using equipment to measure how the waves bounced back to the receiver on Earth, scientists made a rough map of a portion of the planet's surface. The map gave scientists some idea of the surface features on Venus.

Venus remained hidden beneath its atmosphere of clouds. While astronomers on Earth continued to probe the planet using radar, there were many questions about Venus that could be answered only by going there. Beginning in the early 1960s, a number of spacecraft traveled to the planet to look deeper into the mysteries of Venus.

# Venus

### Age
About 4 billion years

### Diameter
7,503 miles (12,070 kilometers)

### Distance from the Sun
About 67 million miles (108 million kilometers)

### Closest approach to Earth
25 million miles (40 million kilometers)

### Orbital period (year)
225 Earth days

### Rotation period (day)
243 Earth days

### Temperature
900° F (482° C) at surface

### Planetary mass
82 percent of Earth's mass (Earth's mass is about 6,000 million, million, million tons)

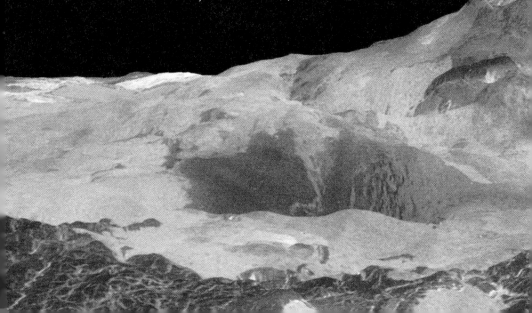

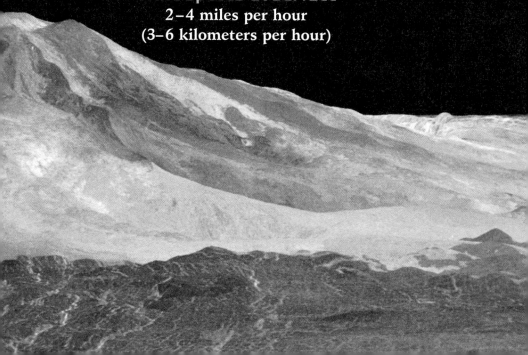

## Composition
Molten metal core; mantle and crust of various rock

## Atmospheric composition
98 percent carbon dioxide; traces of nitrogen and other gases

## Atmospheric pressure at surface
1,300 pounds per square inch (90 times Earth's pressure)

## Surface gravity
91 percent of Earth's gravity

## Inclination of axis ("tilt" of planet's axis)
178 degrees

## Color of sky as seen from surface
Dull orange or yellow

## Color of landforms on surface
Shades of dull yellow and gray

## Wind speeds at surface
2–4 miles per hour
(3–6 kilometers per hour)

# 3

# Visits to Venus

The development of spacecraft in the late 1950s and early 1960s gave astronomers a new way to explore the planets of our solar system. From 1962 to the end of the century, more spacecraft visited Venus than any other planet.

## Mariner 2

The first spacecraft to arrive at Venus was *Mariner 2* in December 1962. *Mariner 2* was important as the first spacecraft ever to explore another planet. Its flyby of Venus lasted only forty-two minutes. But in that short time, the spacecraft's instruments discovered that the cloud tops on Venus were much hotter than anyone had expected. As *Mariner 2* traveled around to the dark side of the planet, it discovered that the clouds on the dark side were just as hot as those on the sunlit side. This discovery was the first strong scientific evidence that Venus was experiencing a runaway greenhouse effect. It was also *Mariner 2* that discovered the planet's tremendous atmospheric pressure and its intense surface temperature of 900° F (482° C).[1]

Mariner 2 *discovered that Venus's surface temperature is as high as 900 degrees Fahrenheit.*

### The Venera Probes

The United States was not the only country launching spacecraft to Venus. The Russian space program built a series of Venera space probes with armor, designed to survive the atmosphere and pressure of Venus. On December 15, 1970, *Venera 7* parachuted through the clouds of Venus, making a soft landing on the planet. *Venera 7* transmitted data back to Earth for about twenty minutes after landing. Then the planet's heat and pressure combined to melt the spacecraft's transmitter.

*Venera 8* landed on Venus in 1972, sending data to Earth for over an hour. Both *Venera 7* and *Venera 8* gave scientists information about the planet's temperature, pressure, and atmospheric components before their signals failed. The scientific equipment aboard the two spacecraft were chilled to very low temperatures before descending through the atmosphere of Venus. They needed to be chilled in order to survive for any length of time in the high heat at the planet's surface.

Scientists made further improvements to make the equipment last longer for the next lander, *Venera 9*.[2] In 1975, *Venera 9* landed on Venus and sent the first black-and-white pictures of the planet's surface. The image showed many rocks and bits of gravel. One Russian scientist described the scene in the image as a "stony desert."[3]

Also in 1975, *Venera 10* sent back the first color pictures, showing the rocky landscape and churning yellowish sky of Venus. The images also showed a landscape that looked geologically much older than the one at the landing site of *Venera 9*. This suggested that parts of the planet had experienced volcanic activity or some other movement of surface rock in the past few million years.

## Mariner IO

In 1974, NASA's *Mariner 10* passed by Venus on its way to Mercury. It recorded many images of the planet's upper atmosphere and made further studies of the very small temperature differences between the daytime and nighttime sides of the planet.[4] *Mariner 10*'s temperature data helped confirm earlier theories of a powerful greenhouse effect on Venus.

Despite the success of the Venera series and NASA spacecraft like *Mariner 10*, there was still little known about the surface of Venus. Were there craters? Were there mountains? Did the sulfuric acid in the atmosphere come from erupting volcanoes on the planet? To answer these and other questions, scientists needed a spacecraft that could make a detailed map of the planet's surface.

## Pioneer Venus Orbiter

In May 1978, NASA launched the *Pioneer Venus Orbiter* to Venus. It was supposed to orbit the planet for two years, using radar signals to produce a map of the surface. *Pioneer* worked far beyond scientists' expectations. It orbited Venus for fourteen years, making a map of the planet that revealed many mountains, plains, and valleys. But more questions remained.

Had the valleys been created by water long ago? Was the surface being shaped by volcanic activity? What was the surface made of? And how old was it? Only a more detailed map of the planet's surface could help answer those questions.

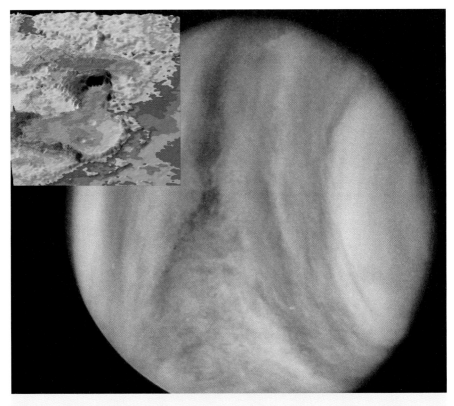

*The* Pioneer Venus Orbiter *mapped Venus for fourteen years. The probe took the background image of Venus in 1979. It also provided an image of Ishtar Terra (inset), a plateau on Venus that is as large as Australia. The greens and blues show low elevations. The reds and yellows are higher areas.*

### Magellan

The spacecraft *Magellan* was launched from the space shuttle *Atlantis* in May 1989. After two complete orbits around the Sun, *Magellan* arrived at Venus on August 10, 1990. Using cloud-penetrating radar beams, *Magellan* spent the next four years mapping 98 percent of the planet's surface in greater detail than ever before.

The amount of data the spacecraft sent back was impressive. By the end of its first year of mapping, *Magellan* had completed more than two thousand orbits. It had traveled 75 million miles around the planet, and had led to remarkable discoveries.

### Planet of Volcanoes

Volcanoes of all shapes and sizes pockmark the surface of Venus. New lava is spread across high areas. The *Magellan* images showed scientists that volcanic activity played an important part in forming the landscape of Venus. The flow of lava from volcanoes was changing the landscape of Venus just as water erosion changes the landscape of Earth.[5]

Many of the long, hardened lava flows on Venus show large cracks. The cracks were caused by quakes that occurred when giant pieces of the planet's surface slightly shifted their position. On Earth, larger pieces of Earth's crust sometimes move or shift, causing earthquakes.

In some places on Venus, scientists found so many

lava flows that it was difficult to separate one from another. Where the flows came together, it was sometimes almost impossible to see which of the overlaid lava flows was older and which was younger. Only a very close and careful study of *Magellan's* images could identify the different ages of the overlaid lava flows.

Scientist Michael Lancaster helped other scientists study a series of lava-flow images. One set of images was from a volcanic area on Venus called Mylitta Fluctus.

*Lava flows on Venus created its unusual landscape. The dome-like hills may be the result of thick lava eruptions. The color enhancement of this photo was based on color images from the Venera spacecraft.*

Sorting out the complicated glob of lava flows became frustrating at times.

"Everyone says Olympus Mons on Mars is the biggest volcano in the solar system," Lancaster joked one day. "It isn't. Venus is. The entire planet is one big volcano."[6]

### Channels on the Surface

Volcanoes and quake fractures were not the only things scientists discovered from *Magellan*'s radar images. The spacecraft mapped a number of channels running along the planet's surface. Scientists believed they had been cut by hot lava flows. But one of the channels was of incredible size.

In some places, the channel was more than two miles wide—as wide as thirty-five football fields placed end to end. Scientists studying *Magellan*'s images traced the course of the channel across a plains region of Venus for more than 4,200 miles. That distance made the channel longer than the entire Nile River, the longest river on Earth.

"It follows a . . . smoothly-curving course," said *Magellan* project scientist Dr. Saunders, "that can be traced continuously on the surface." The channel is the longest channel known in the entire solar system.[7]

"The very existence of such a long channel is a great puzzle," Saunders added. "If the long channel were carved by something flowing on the surface, the liquid must have had unusual properties."[8]

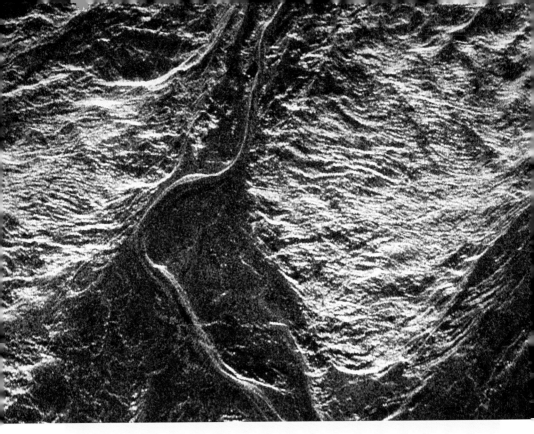

*Volcanic channels on Venus have been cut by hot lava flowing on the surface of the planet.*

At first, Saunders and other scientists did not think even lava from a volcano could have created such a long channel.

"Lava, even very high temperature types, would need to have a very high rate of flow to go so far," Saunders said. He and other scientists required further studies of the *Magellan* images before they could conclude that lava had created the long channel.

"The challenge of understanding the origin of this channel will lead to better understanding of planetary geological processes," Saunders said.[9]

### Craters Cover Venus

The *Magellan* images also revealed that the planet's surface is covered with many craters from meteor and asteroid impacts. The spacecraft discovered nearly nine hundred craters, ranging in size from one mile wide to nearly two hundred miles wide. Some of the craters had impacted the surface in strange patterns that scientists had never seen before. The dense atmosphere of Venus caused these strange cratering patterns.[10]

Scientists had studied images of craters from the Moon and from Mercury and Mars. While there were many similarities between those craters, they were different from the craters and impact patterns found on Venus.

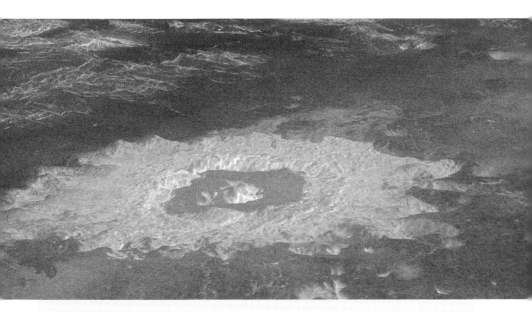

*There are many impact craters on the surface of Venus. They are scattered all over the planet. The spacecraft* Magellan *observed these three craters.*

Atmospheres cause smaller meteors to be burned up by friction before hitting the surface. The very thin atmosphere of Mars burns up some very small meteors but has little or no effect on larger objects from space.

"We haven't thought much about the effects of atmospheres on cratering," said scientist John Guest, "and on Venus we have the most extreme atmosphere of all."[11]

The atmosphere of Venus is thick enough, scientists believe, to cause large meteors to heat up and explode into smaller pieces as they enter the atmosphere. The impact of these many pieces creates the unusual cratering patterns *Magellan* revealed on the surface of the planet.

## A Changing Landscape

Venus's atmosphere may protect the planet's surface from many meteors, but the surface is still changing. *Magellan*'s many passes over the same areas of the planet showed that a giant landslide had taken place since the beginning of the spacecraft's mission. Scientists compared previous mapping data with more recent images from the spacecraft. They noticed that a three-mile cliff on one part of the planet had been reduced to a fan-shaped area of rubble.

The landslide images were more proof that the surface of Venus is changing. Evidence of quakes and volcanoes is everywhere. Small mountains, called coronae, were created when a volcanic dome rose up. When the lava beneath it flowed away, the dome collapsed on itself before it could cool and become solid. Flattened "pancake domes" formed when lava oozed out onto the flat surface, spreading out in every direction for miles, until the entire mass cooled and hardened.[12]

*Magellan*'s radar images of Venus's surface gave scientists a wealth of new information that would be studied for many years. Its detailed map of the planet would help scientists discover clues about Venus's past and make predictions about its future.

*Magellan* sent data to Earth until October 1994, when the spacecraft's solar panels began to fail. It remained in

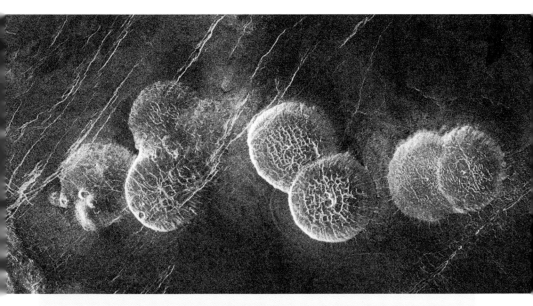

*Pancake domes formed on Venus when the lava flowed out onto the flat surface of the planet. Then the lava cooled and hardened.*

orbit around Venus for another year, while NASA used the spacecraft to test new orbit-changing maneuvers for future spacecraft. It finally burned up in the atmosphere of Venus in October 1995.

## Future Missions to Venus

NASA is considering two future missions to Venus between 2000 and 2010. The Venus Multiprobe Mission plans to land sixteen miniature probes on the planet in order to measure its atmospheric properties in more detail at different locations on the surface. The second mission, the Venus Composition Probe, will study how much water existed in the planet's past.

"We've had a look at how Venus has operated over the last few hundred million years," said scientist James Head, "but we don't know what happened in the previous 90 percent of the planet's history. The same is true of our knowledge of Earth, where we know only about the last 20 percent of its history."

"It's important not to wipe our hands and walk away from these questions," he added. "It's important to see what the threads are today that lead back to the early days, both here and on Venus."[13]

While we wait to learn more about our nearest planet, you may want to make a study of Venus yourself, by looking for it in the sky. First, you will have to know when and where to look.

# 4

# The Brightest Planet

Go outside on a clear night right after sunset, and look to the west. The following morning, go outside just before sunrise, and look to the east. If you see a very bright object in the sky at either of these times or places, chances are that it is Venus.

Venus, however, is not always visible. Venus orbits close to the Sun, so the planet always appears close to the Sun in our sky. When it passes in front of or behind the Sun, it is invisible to us, lost in the Sun's daytime brightness. Only when Venus's orbit has carried it far enough outward from either side of the Sun can it be seen easily from Earth.

At these times in Venus's orbit, the planet will be visible only in the western sky or near the western horizon during the hours of dusk, or in the eastern sky near the eastern horizon in the hours before dawn. Anyone curious about viewing Venus can check current astronomy magazines or various astronomy Web sites on the Internet to learn what time and in what part of the sky Venus will be visible.[1]

To the naked eye, Venus will be the brightest object in the sky, unless there is a full Moon. Binoculars or a telescope will reveal the planet's shape, which may be in a phase, depending on where the planet is in its orbit when you are observing it.

Observing Venus from Earth is much more pleasant than it would be to observe it from the planet's surface. But a visitor could learn a great deal more about the planet by being there. Such a visit would not be easy.

### A Visit to Venus?

Even if a spacecraft could survive the descent through Venus's atmosphere, stepping outside of the spacecraft would be a very difficult task. Visiting astronauts would need very special spacesuits to keep from being crushed by the planet's intense atmospheric pressure. Being able to walk and breathe under such heavy pressure would require a slightly better spacesuit than we are able to make with today's technology.

If such a suit existed, astronauts could walk along the rocky surface and look up into the yellowish sky of churning clouds. The winds across the surface would make walking very strange. Because the air is so dense near the surface of the planet, even a four-mile-per-hour wind would make the astronauts feel as if they were wading through water and being struck by waves.

There would be light in the daytime, but the astronauts would never see the disk of the Sun. As the human visitors looked out into the distance, the landscape would be scattered with rocks, gravel, and flattened boulders. Depending on where they landed, they might see a mountain in the distance, the raised rim of a crater, or the ledge of a deep channel.[2]

The visiting astronauts might explore what some scientists believe is an area of fresh lava flows around the six-mile-high volcano, Maat Mons. This volcano may have spewed millions of tons of sulfur dioxide into Venus's atmosphere as recently as fifty years ago.

*Venus may be visible in the early evening (western sky) or at dawn (eastern sky). You can use a pair of binoculars to see it.*

As the astronauts explore the volcano, they might be caught in rainfall. On Venus, as a result of such volcanic eruptions, the sulfur dioxide has been falling from the sky in the form of sulfuric-acid rain. The astronauts' spacesuits had better be acid-proof! Without such protection, these raindrops would burn holes in their suits, as well as their bodies.

The gravity on Venus is about the same as on Earth, so an astronaut will weigh about the same. A person who weighs one hundred pounds on Earth would weigh ninety-one pounds on Venus. Although their weight would feel about the same, the atmospheric pressure would exert tremendous stress on their bodies.

The astronauts would also need to bring along a tank of air to breathe. All of Venus's air is poisonous.[3]

Despite these dangerous conditions, humans may someday find a way to visit the planet Venus. Perhaps then we will be able to solve more of its many mysteries.

Scientists' efforts to understand Venus through the use of telescopes and spacecraft have been important to our understanding of the processes at work on our own planet. Many scientists believe it is possible that Earth may be slowly experiencing the same processes that caused Venus to become the hot and hostile place that it is today. Further studies of our nearest planet may teach us about the future of our own planet.

While Venus may be burning hot and choked with poisonous gases, it has been a bright beacon in Earth's sky through all of humanity's existence. Today we do not offer blood sacrifices to the planet as did the ancient Maya, nor do we believe that Venus is really a goddess of love. It is simply another planet, the nearest one to Earth, orbiting the Sun.

But it is not a simple planet. It still holds many mysteries that scientists hope to solve. Until then, we can look with curiosity and wonder to that bright point of light in the morning or evening sky, that is the planet Venus.

*The* Magellan *spacecraft, shown here being released from the space shuttle* Atlantis*, helped scientists learn many new things about Venus. Future missions to the planet will yield even more exciting discoveries about Earth's sister planet.*

# CHAPTER NOTES

## Chapter 1. Peeking Through the Clouds

1. NASA Jet Propulsion Laboratory, *Magellan Press Releases,* "*Magellan* Maps 90% of Venus," July 26, 1991, <http://www.jpl.nasa.gov/magellan/pr1383.html> (February 15, 1999).

2. NASA Press Releases, "New *Magellan* Global Views of Venus Released," March 16, 1995, <http://www.qadas.com/qadas/nasa/nasa-hm/0082.html> (November 3, 1999).

## Chapter 2. Goddess of Love, Planet of Heat

1. Terence Dickinson, *Nightwatch: An Equinox Guide to Viewing the Universe* (Ontario, Canada: Firefly Books, 1996), p. 114.

2. Peter Cattermole and Patrick Moore, *Atlas of Venus* (New York: Cambridge University Press, 1997), pp. 1–4.

3. Jean Audouze and Guy Israël, eds. *The Cambridge Atlas of Astronomy,* 3rd ed. (New York: Cambridge University Press, 1994), p. 81.

4. Nicholas Booth, *Exploring the Solar System* (New York: Cambridge University Press, 1996), p. 80.

5. Audouze and Israël, p. 78.

6. Thomas R. Watters, *Planets: A Smithsonian Guide* (New York: Macmillan Publishing Company, 1995), p. 60.

## Venus Table

Jean Audouze and Guy Israël, eds. *The Cambridge Atlas of Astronomy,* 3rd ed. (New York: Cambridge University Press, 1994), pp. 56, 64, 67, 78–79; Peter Cattermole and Patrick Moore, *Atlas of Venus* (New York: Cambridge University Press, 1997), p. 3.

### Chapter 3. Visits to Venus

1. Nicholas Booth, *Exploring the Solar System* (New York: Cambridge University Press, 1996), p. 70.

2. Peter Cattermole and Patrick Moore, *Atlas of Venus* (New York: Cambridge University Press, 1997), pp. 33–34.

3. Ibid., p. 35.

4. Ibid., p. 34.

5. NASA Jet Propulsion Laboratory, *Magellan Press Releases*, "*Magellan* Maps 90% of Venus," July 26, 1991, <http://www.jpl.nasa.gov/magellan/pr1383.html> (February 15, 1999).

6. Henry S. F. Cooper, Jr., *The Evening Star: Venus Observed* (New York: Farrar Straus Giroux, 1993), p. 180.

7. NASA Jet Propulsion Laboratory, *Magellan Press Releases*, "*Magellan* Discovers Longest Channel in the Solar System," August 30, 1991, <http://www.jpl.nasa.gov/magellan/pr1387.html> (February 15, 1999).

8. Ibid.

9. Ibid.

10. Cattermole and Moore, pp. 80–81.

11. Cooper, p. 86.

12. Thomas R. Watters, *Planets: A Smithsonian Guide* (New York: Macmillan Publishing Company, 1995), pp. 63, 65.

13. Cooper, pp. 262–263.

### Chapter 4. The Brightest Planet

1. Terence Dickinson, *Nightwatch: An Equinox Guide to Viewing the Universe* (Ontario, Canada: Firefly Books, 1996), pp. 112–114.

2. Jean Audouze and Guy Israël, eds. *The Cambridge Atlas of Astronomy*, 3rd ed. (New York: Cambridge University Press, 1994), pp. 84–89.

3. Henry S. F. Cooper, Jr., *The Evening Star: Venus Observed* (New York: Farrar Straus Giroux, 1993), p. 22.

# GLOSSARY

**electromagnetic spectrum**—The entire range of energy. The energy, or radiation, comes to us in waves of electricity and magnetism. The light that we see is only one of seven kinds of radiation and is near the middle of the spectrum.

**greenhouse effect**—The warming of a planet through the trapping of heat by the atmosphere. Carbon dioxide in Venus's atmosphere allows the Sun's heat to penetrate to the planet's surface but does not allow all of it to escape, causing the heat to build up.

**lava**—Liquid rock that rises to the surface from a volcano.

**magnetic north**—The direction that a magnetic needle points due to the planet's magnetic field.

**NASA** (National Aeronautics and Space Administration)—The United States government agency in charge of space activities.

**plateau**—A large area of land that has a level surface and is raised above the surrounding land on at least one side.

**radar**—A radio system that uses ultrahigh-frequency radio waves, reflects them off an object, then analyzes the returning radio waves to determine characteristics such as the size or distance of the object.

**sulfuric acid**—A highly corrosive acid. Clouds of sulfuric acid exist on Venus, and the acid falls to the surface as rain.

tectonism—The process in which large pieces of a planet's crust move, fold, and shift to create the planet's surface terrain.

ultraviolet light—A kind of radiation in the electromagnetic spectrum that is invisible to human eyes. Ultraviolet rays have shorter wavelengths than visible light and can cause blindness and sunburn.

volcanism—The process of molten rock pushing up through the planetary crust to vent itself out onto the surface. This venting causes mountains or hills to form at the site of the vent.

# FURTHER READING

## Books

Branley, Franklyn M. *Venus: Magellan Explores Our Twin Planet.* New York: HarperCollins, 1994.

Kipp, Steven L. *Venus.* Manakato, Minn.: Bridgestone Books, 1998.

Vogt, Gregory L. *Venus.* Brookfield, Conn.: Millbrook Press, 1994.

## Internet Addresses

Arnett, Bill. "Venus." *The Nine Planets: A Multimedia Tour of the Solar System.* April 27, 1999. <http://seds.lpl.arizona.edu/billa/tnp/venus.html> (May 25, 2000).

Baalke, Ron. "*Magellan* Mission to Venus." *Magellan Project Homepage.* n.d. <http://www.jpl.nasa.gov/magellan/> (May 25, 2000).

Hamilton, Calvin J. "Venus Introduction." *Views of the Solar System.* © 1997–1999. <http://www.solarviews.com/eng/venus.htm> (May 25, 2000).

The Regents of the University of Michigan. *Windows to the Universe.* © 1995–1999, 2000. <http://www.windows.umich.edu> (May 25, 2000).

# INDEX